Table of Contents

1. INTRODUCTION

The advance of Web services technologies promises to have far-reaching effects on the Internet and enterprise networks. Web services based on the eXtensible Markup Language (XML), Simple Object Access Protocol (SOAP), and related open standards, and deployed in Service Oriented Architectures (SOA) allow data and applications to interact without human intervention through dynamic and ad hoc connections. Web services technology can be implemented in a wide variety of architectures, can co-exist with other technologies and software design approaches, and can be adopted in an evolutionary manner without requiring major transformations to legacy applications and databases.

The security challenges presented by the Web services approach are formidable. Many of the features that make Web services attractive, including greater accessibility of data, dynamic application-to-application connections, and relative autonomy (lack of human intervention) are at odds with traditional security models and controls. The following are some of the challenges for secure web services:

- Confidentiality and integrity of data that is transmitted via Web services protocols in service-to-service transactions, including data that traverses intermediary (pass-through) services.

- Functional integrity of the Web services that requires both establishment in advance of the trustworthiness of services in orchestrations or choreographies.

- Availability in the face of denial of service attacks that exploit vulnerabilities unique to Web service technologies, especially targeting core services, such as discovery service, on which other services rely.

The SOA processing model requires the ability to secure SOAP messages and XML documents as they are forwarded along potentially long and complex chains of consumer, provider, and intermediary services. The nature of Web services processing makes those services subject to unique attacks, as well as variations on familiar attacks targeting Web servers.

In web services, the service-level compositional techniques create complex inter-dependencies between services belonging to different organizations that can be exploited due to some localized or compositional flaws. Therefore such exploits/attacks [1-3] can affect multiple servers and organizations, resulting in financial loss or infrastructural damage. Investigating such incidents requires that dependencies between service invocations be retained in a participating party neutral and secure way. Material evidence currently extractable from web servers such as log records, firewall alerts from end point services, and the like, do not have forensic value because defendants can claim that they did not send that message. In this report, we describe a participant neutral solution for a forensically valid evidence gathering mechanism for web services.

2. BACKGROUND ON WEB SERVICES

Two conceptual elements underlie current web services: (1) Use of XML (eXtensible Markup Language), SOAP (Simple Object Access Protocol), and WSDL (Web Service Definition

Language) as basic building material; (2) Complex applications built upon long-running transactions that are composed of other web services.

2.1 Basic Appearance of Web Services

XML format underlies the entire web service architecture and its artifacts. All schemas, definition files, messages transmitted are formed by the means of XML. WSDL, a XML based definition file, defines the interface of a web service in order for the service to be invoked by other services in accordance with the specifications of internal executions. SOAP, a XML based protocol, defines the metadata of the messages to be exchanged between services. WSDL documents define operations; and they are the only mechanisms in order for web services to communicate with each other. Web services use SOAP messages by exchanging them as incoming and outgoing messages through the operations.

2.2 Composition of Web Services

The message exchange patterns (MEP) described above constitute the entire web service paradigm. These simple MEPs construct collaboration scenarios using the appropriate composition models. While defining a composition, two issues come up: how it is designed and what pattern it employs.

2.3 Static vs. Dynamic Composition

One consumer service could select the target provider service either statically or dynamically, that is, at design-time or run-time. Design-time selections entail a-priori determination while run-time selections can introduce the opportunity to switch between web services dynamically. Static web service composition introduces less anonymity than the dynamic counterpart, therefore it requires less effort for a forensic examination.

2.4 Hierarchical vs. Conversational Composition

According to Khalaf [4], one could compose web services either in a hierarchical or in a conversational pattern. Through *Hierarchical Composition*, a consumer web service calls another composite web service passing the input parameter and receiving the result. Other than this request-response activity, no other call is employed to the same instance at the target. Designers, however, mostly use *Conversational Composition* when web services need to interact with each other more than once throughout the same instances at both sides. In these scenarios, the target system, unavoidably, makes its internal state mutable, thus causing overlapping instances to be created within parties to the composition.

In a hierarchical pattern, the instance of an external web service completely finishes before returning the result while many interactions between instances can survive in conversational pattern. Although describing what exactly happened during execution in a hierarchical pattern is reasonable, this might not be the case in conversational pattern. Thus, from a forensics point of view, representing and recreating the activities in the latter pattern is much more difficult than the former one.

2.5 Composition Standards and Languages

Although there are many standards and specifications for web services, we mention state-of-the-art orchestration and choreography specifications here. BPEL (Business Process Execution Language) is a language for business process modeling. WS-BPEL and BPEL4WS are their two popular implementations for web service architecture. They can define both abstract and executable processes. They are two effective tools to realize orchestration of composite web services from a central point. Conversely, WSCI and WS-CDL create a global view of multi-party choreographies of web services from their individual description files. These languages enable collaborative processes that are recruiting multiple web services, and facilitate interactions between them from a global, high-level perspective rather than an individual service's request response perspective.

2.6   Web Services Example

Before examining forensics for Web services in detail, it is helpful to first consider an example of Web services that can be used as a model to understand Web services security. Figure 1 shows a simplified representation of major Web services components for a consumer loan service.

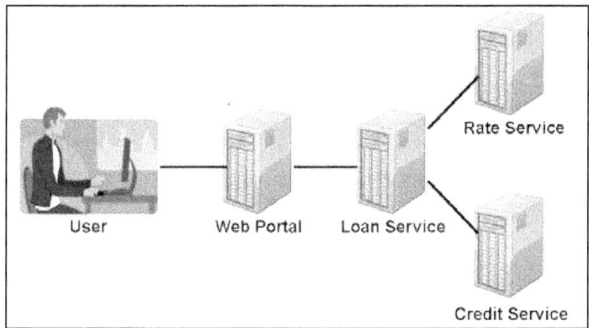

Figure 1: Consumer Loan Service Example

In this example, users (consumers) contact a Web portal that offers financial services. When a user requests loan information, the Web portal contacts a loan Web service on behalf of the user. The loan service then contacts other Web services as needed, such as rate and credit services, to get up-to-date information, and passes the requested information back to the user through the

Web portal. Each host that needs to use a Web service uses the Universal Description, Discovery, and Integration (UDDI) protocol to locate a suitable Web service and invoke it.

There are several major security objectives for this scenario. One is validating the identity of the user requesting the loan information. Another is restricting the use of Web services; for example, the credit service might charge a transaction fee for each request, so the credit service would need assurance of the identity of the service requesting the information. Other security objectives involve ensuring that only authorized users and Web services are able to access, modify, and/or delete the necessary information.

3. WEB SERVICE ATTACKS

There are many attacks on web services, such as WSDL/UDDI scanning, parameter tampering, replays, XML rewriting, man-in-the-middle, eavesdropping, routing detours [1-3] etc. In addition to web service attacks classified in [1], dynamic service selection, choreography, orchestration, and composition increase the ways of exploiting web services, such as application and dataflow attacks [3].

We now show the details of a sample cross-site scripting (CSS) attack used to illustrate the capabilities of FWS. A typical CSS attack may inject a malicious script to harm a web service that dynamically builds some of its information. Figure 2 shows an attacker with stolen credentials injecting some malicious data invoking an update operation on a Weather service that stores this script (including instructions to steal cookies) from web browsers. Then a web application, say *Portal Web Application*, invoking a GET operation retrieves this malicious data and publishes the weather information to its subscribers in an **html** form, thereby making the subscribers send their personal information stored in cookies to the attacker's *Fishing Net Application*. Then, a *Fishing Net Application* managed by the *Attacker* can obtain sensitive user information as shown in Figure 2.

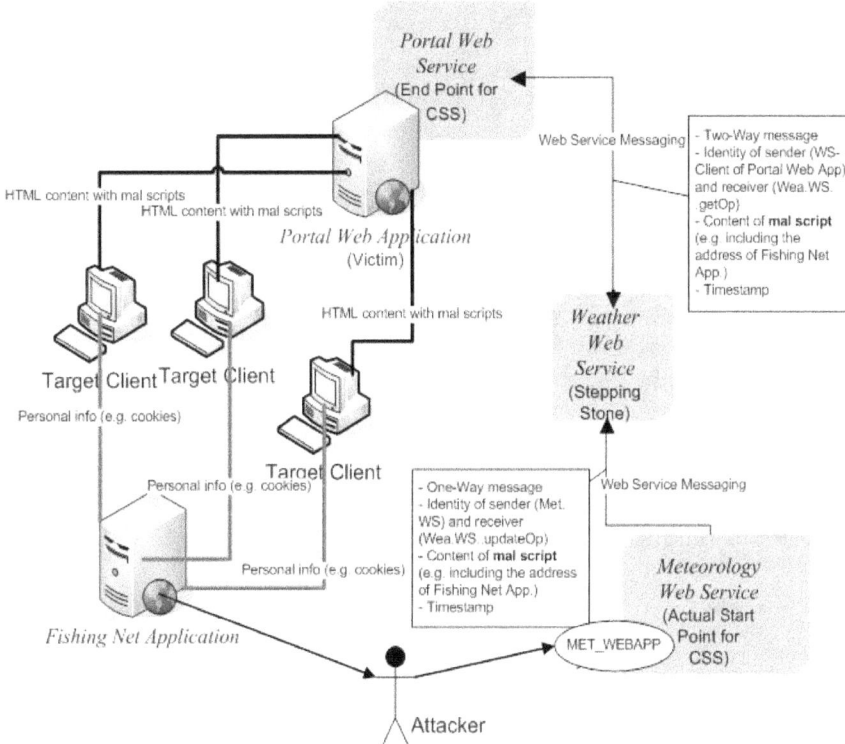

Figure 2. A Cross-Site Scripting Attack Using Web Services. Meteorolgy Web Service (MET_WS) gets infected with malicious data and delivers the data ignorantly to the Weather Web Service (WEA_WS) when requested. WEA_WS, accordingly to their choreography, passes malicious data to Portal Web Service (POR_WS) among other legal information. An attacker, aware of choreography among web services, exploits this model and has Portal Web Application delivered malicious data to its members using web services in this choreography model.

The stated CSS attack shows how the business logic of a web service can be used to attack a server that depends upon other web services. In this scenario, *Portal Web Service* can claim that *Weather Web Service* sent the malicious content, whereas the actual source was *Meteorology Web Service*. This illustrates the need to have a mechanism that irrefutably points to the source of malice.

4. CHALLENGES IN FORENSICS OF WEB SERVICES

As opposed to traditional forensics implementations, applying forensics to web service infrastructures introduces novel problems such as a need for neutrality and comprehensiveness. The primary purpose of digital forensics is to present digital evidence in legal proceedings. Therefore, the techniques used to extract digital evidence from devices must comply with legal standards. Reliability is another important issue for forensic examinations.

a) Neutrality

Web services, owned by organizations, have equal rights in the court of law when any dispute arises between parties. Any log records residing at only one party's site would have no forensic value under these circumstances since any alteration on the records might have been employed in favor of that site. Many forensic investigations on traditional systems have been based on one site's records. For traditional systems that may be reasonable since investigators take advantage of inquiring users and human factors to corroborate evidence. In service oriented architectures, both sites should be automated to collect and retain their own records. Records at both the sites would be under question by the opponent party, thus showing the need to have a neutral party capturing and preserving evidence based on interactions between parties.

b) Comprehensiveness

As described earlier web service compositions may span over many web services of many organizations. Such interdependent services create long information flows. Thus malicious data may stream over many web services. From a forensics point of view, the evidence gathered should be comprehensive enough so that investigation can reach all related end points to reveal what party performed what action. Incomplete evidence might point to any web service node as the source of malicious activity, thus misleading the investigators through the examination.

c) Reliability

Yet another important principle that any evidence should have is reliability. In the court of law, digital evidence must be presented in an articulate manner. Because impersonation and replay attacks do occur in web services, cryptographic mechanisms would help to protect ownership of information passed around in messages by signing them digitally. Such a requirement, of course, would entail web services using a cryptography platform such as Public Key Infrastructure (PKI).

5. OVERVIEW of FWS

In order to facilitate and base forensic investigations on reliable data, we propose designing *Forensic Web Services (FWS)* that preserve appropriate evidence to recreate the composed web service invocations. This would have a greater chance of being accepted in a court of law. FWS will provide on-line forensic capabilities to other web services as a web service itself. To utilize them, FWS would be integrated with web services that require their services – refered to as customer web services of FWS. In order to do so, FWS provides a centralized service access point to its customer web services. This information retained by FWS acting as a trusted third party can be directly given to forensic examiners. Previous proposals to monitor web services [5] and generating evidence [6-8] have been for business purposes. The evidence they produce does not meet the requirement for forensic examinations.

The Forensic Web Service Framework provides two essential services:

1. **Pair-wise evidence generation:** Collect transactional evidence that occur between pairs of services at service invocation times. Figure 3 illustrates this process, called "deliver".

2. **Comprehensive evidence generation:** On demand, compose pairs of transactional evidence collected at service invocation times and reveal global views of complex transactional scenarios that occurred during specified periods, and provide them for forensic examiners. Figure 3 illustrates the 'collectDependents' algorithm (the core component of this process), that is inspired by King's and Chen's dependency graph algorithm [9].

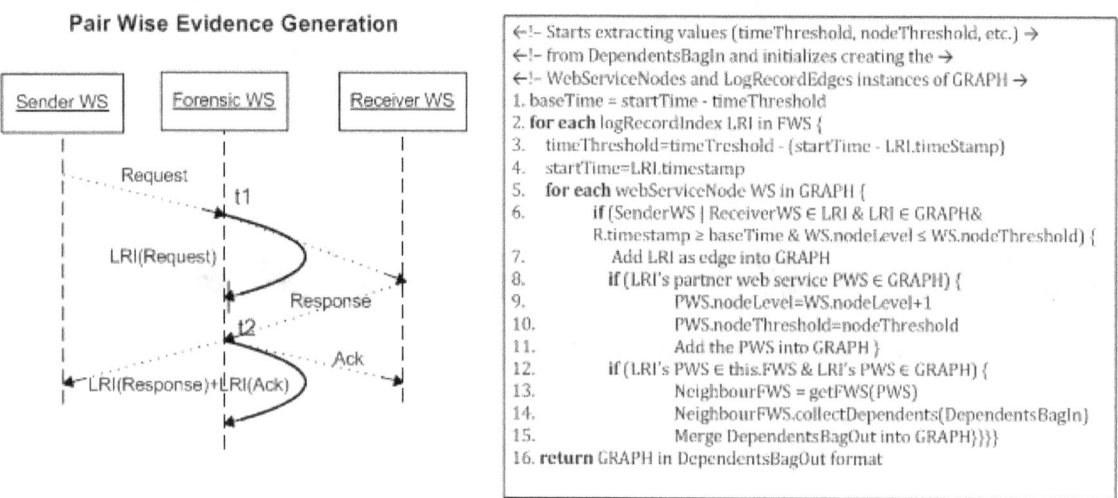

Figure 3. Pair-wise evidence generation.

5.1 FUNCTIONS PROVIDED BY FWS

Organizations that are tightly integrated with each other through web transactions and processes can benefit from FWS in many ways. First, organizations need to hold some of their partner web services accountable when their mal-actions affect one's own efficiency, consistency, availability, etc. Secondly, the detailed explanation of the malicious activity may impact the severity of punishments or collectible monetary compensation. Logging of critical information exchanges is an effective way to meet these two needs. FWS can monitor the systems nonrefutably; those records retained by the system would have forensic value in a court of law; which has not been the case so far. Hereafter, we propose to provide more refined evidence regarding the activities that occur on web service architectures. In the next section, we propose an architecture of the system to maintain instance correlations through both hierarchical and conversational compositions.

5.2 Monitoring Web Services Interactions

Through the FWS framework, WS-Forensics layer (see Figure 4) routes the interactions to pass over FWS stations on the way to their ultimate targets. As described in our previous study [10], handler-chain architecture [11] is used to ease and standardize client side workload on deployment of FWS-Handlers. This function underlies the entire forensic functionality of the FWS described below.

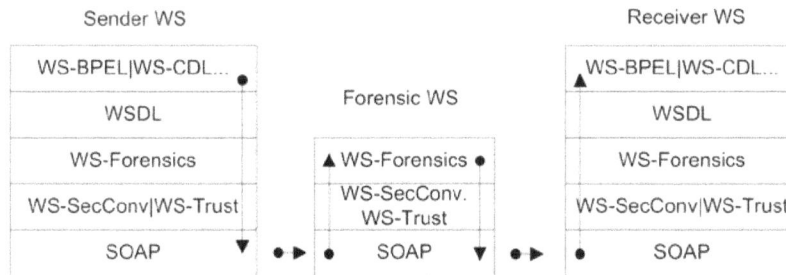

Figure 4. WS-Forensics Stack. *Arrows depict how WS-Forensics is applied for a message through web services and their existing stacks.*

6. Forensics over Web Services

Capturing the interdependent activity makes little sense from a forensics perspective if the capturing procedure is not comprehensive. Finding the dependent interactions and web services with respect to a specific point in the scope of a certain composed execution of web services seems an exhaustive task. In this section, we give an overview of the architecture of FWS.

We propose a protocol in order for FWS-Handlers and FWS stations to run in the layer proposed above. FWS stations store interactions to ease the task that should be performed by the algorithms (see [10] for design) to collect records of dependent interactions spanned over many web services during the actual execution. Figure 5 illustrates typical message flows for forensics capabilities for web services.

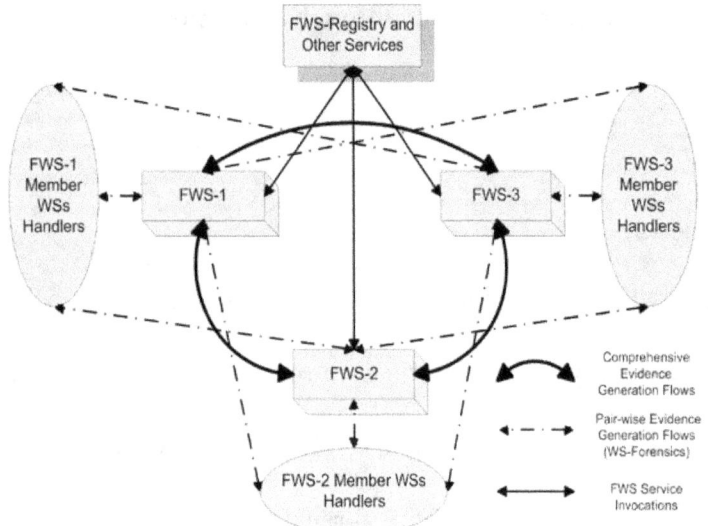

Figure 5. The FWS Framework and Message Flows.

Revealing Global Composition Instances

Many studies/specifications offer composition models to handle business transactions and other cross organizational activities over web services. To the best of our knowledge, there is no existing framework that can create the sceanario of interactions among the services from the events logged in a neutral way. FWS can interleave the instances of global / composed executions of web services using global unique identifier as shown in Figure 6-I. We believe that with our design, it should be possible to reveal and represent the composition of executions. This capability would be provided on the basis of the following two functions; verification of orchestration processes and choreography instances.

Orchestration Process Verification

Given an orchestration process model of a web service, the FWS framework can detect whether the process behaved as it is expected. When such checks are applied to web services based on the instances revealed above, the results can determine if a deviation has occurred from the expected behavior of the services.

Choreography Instance Verification

Given a choreography model, FWS could detect deviations from expected set of choreography instances and represent them as shown in Figure 6-II. Deviating points in the choreography instance should successfully be addressed along with actual identities of sources for deviations to realize any forensic examiner's ultimate goal.

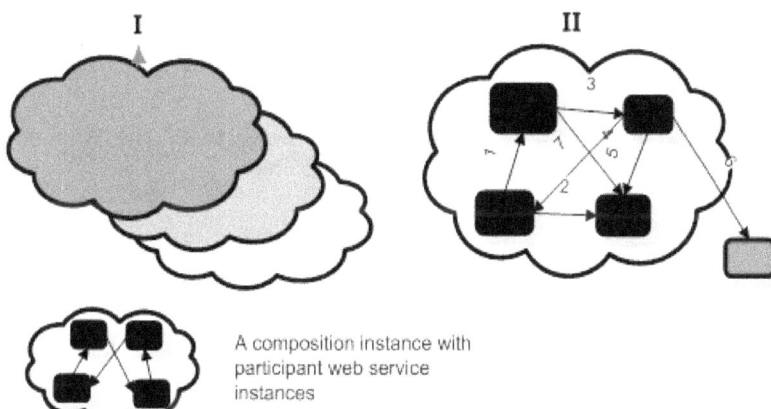

Figure 6. (I) Interleaving the Global Composition Instances. FWS records envision keeps a "global unique identifier" that refers to each separate execution. (II) A Choreography Instance Deviated from Original Model. FWS records are designed to keep dependency information along with instance correlation information thus allowing to reveal if there is any deviation from expected instance of global execution.

A composition instance with participant web service instances

7. RELATED WORK

There is no forensic framework for investigating inter-related web services designed so far. However, the work cited hereafter share some common features with FWS' objectives or methods.

Robinson [12] influenced the model employed through FWS for pair-wise evidence generation with some differences. Robinson [12] provides a framework to support fair B2B communications on the basis of a trusted deliver agent notion. It implements Coeffey-Saidha [13] protocol to provide non-repudiation in their protocols. However, the framework is designed to run with other protocols as well. Robinson[12] only proposes delivering evidences to the related parties, but not preserving them in trusted agents.

Herzberg [6] introduces the notion of having an Evidence Layer for e-commerce transactions. They propose this layer to be at the bottom of the e-commerce stack and on top of a transport layer (such TLS/SSL, or TCP/IP). They introduce two protocols to generate and deliver the evidence to involved parties in message exchange; the first is the *Simple Evidence Layer Protocol* and the second is the *optimistic* one. They employ notaries in the first protocol while generating and delivering the evidence. FWS use the layering approach of Herzberg [6] in the web service stack with minor changes, such as adding the time stamping point, and use their SELP as the pair-wise evidence generation protocol. Like others, Herzberg et al [6] was not designed for forensics.

FWS also implement trusted third parties for pair-wise evidence generation as Coffey-Saidha et al [13]. Although inline TTPs are immature for business transaction, they add value to forensics evidence. Onieva [14] gives the intermediary usage perspective in the implementation of inline TTPs for e-commerce transactions. They also support multi-recipient cases through these intermediaries, but not for forensics. Bilal [15] uses BPEL for non-repudiation protocol implementation in web services, but does not use TTP, thereby lacking the capability to handle message content.

WSLogA [16] track web service invocations by logging service invocations using SOAP intermediaries. Therefore, it captures the external behavior of service invocations. The main purpose of WSLogA is to provide feedback to business organizations by comprehensively logging services usage records. However, because it does not address any distributed collection mechanism necessary to gather comprehensive forensic evidence over services sharing multiple servers.

FWS has been influenced by many studies on network forensics, of which we describe two. Wang uses IDS alerts [17] to generate an evidence graph for network forensic analysis. *Local reasoning* and *global reasoning* help them in defining malicious activity in individual hosts and networks respectively. Unlike Web Server Nodes in FWS, they use hosts as nodes in their graphs.

ForNet [18] is another distributed forensic framework that uses logs from routers in a network to run agents that provide their log records to ForNet servers. Unlike Wang [17], ForNet uses succinct information of every regular network packets adequate to trace the actual source of packets even when they are spoofed. Although not designed for Web Services, this work has been inspired by the design of ForNET.

8. CONCLUSION

Web services span many applications and domains. Consequently, any vulnerability in one service can be exploited to affect more than one service. In Web Services architecture it is a challenge to investigate the nature and source of an attack. We propose a framework referred to as Forensic Web Services that provides this capability as a service to other web services by logging service invocations. Our design shows how collected logs can provide the capability to produce a collection of digital evidence to expose the attack from its logs.

REFERENCES

[1]   A. Vorobiev and H. Jun, "Security Attack Ontology for Web Services," in *Semantics, Knowledge and Grid, 2006. SKG '06. Second International Conference on*, 2006, pp. 42-42.

[2]   Y. Demchenko, L. Gommans, C. de Laat, and B. Oudenaarde, "Web services and grid security vulnerabilities and threats analysis and model," in *Grid Computing, 2005. The 6th IEEE/ACM International Workshop on*, 2005, p. 6 pp.

[3]   A. Singhal, T. Winograd, and K. Scarforne, NIST Special Publication 800-95, "Guide to Secure Web Services" , August 2007, http://csrc.nist.gov/publications/nistpubs/800-95/SP800-95.pdf.

[4]   R. Khalaf, N. Mukhi, and S. Weerawarana, "Service-Oriented Composition in BPEL4WS," in *Twelfth International World Wide Web Conference*, Budapest, Hungary, 2003.

[5]   S. M. S. Cruz, M. L. M. Campos, P. F. Pires, and L. M. Campos, "Monitoring e-business Web services usage through a log based architecture," in *Web Services, 2004. Proceedings. IEEE International Conference on*, 2004, pp. 61-69.

[6]   A. Herzberg and I. Yoffe, "The Delivery and Evidences Layer," Cryptology ePrint Archive Report 2007/139, 2007.

[7]   S. Kremer, O. Markowitch, and J. Zhou, "An Intensive Survey of Non-repudiation protocols," *Computer Communications,* vol. 25, pp. 1606-1621, 2002.

[8]   P. Robinson, N. Cook, and S. Shrivastava, "Implementing fair non-repudiable interactions with Web services," in *EDOC Enterprise Computing Conference, 2005 Ninth IEEE International*, 2005, pp. 195-206.

[9]   S. T. King and P. M. Chen. "Backtracking Intrusions," in *Proceedings of the 2003 Symposium on Operating Systems Principles (SOSP)*, pages 223-236, October 2003.

[10]   M. Gunestas, D. Wijesekera, and A. Singhal, "Forensic Web Services," in *Fourth Annual IFIP WG 11.9 International Conference on Digital Forensics* Kyoto, Japan, 2008.

[11]   P. Srinath, H. Chathura, E. Jaliya, C. Eran, R. Ajith, J. Deepal, W. Sanjiva, and D. Glen, "Axis2, Middleware for Next Generation Web Services," in *Web Services, 2006. ICWS '06. International Conference on*, 2006, pp. 833-840.

[12]   P. Robinson, N. Cook, and S. Shrivastava, "Implementing fair non-repudiable interactions with Web services," in *EDOC Enterprise Computing Conference, 2005 Ninth IEEE International*, 2005, pp. 195-206.

[13]   T. Coffey, P. Saidha, "Non-repudiation with mandatory proof of receipt," ACMCCR: Computer Communication Review 26.

[14]   J. A. Onieva, Z. Jianying, M. Carbonell, and J. Lopez, "Intermediary non-repudiation protocols," in *E-Commerce, 2003. CEC 2003. IEEE International Conference on*, 2003, pp. 207-214.

[15]   M. Bilal, J. P. Thomas, M. Thomas, and S. Abraham, "Fair BPEL processes transaction using non-repudiation protocols," in *Services Computing, 2005 IEEE International Conference on*, 2005, pp. 337-340 vol.1.

[16]   S. M. S. Cruz, M. L. M. Campos, P. F. Pires, and L. M. Campos, "Monitoring e-business Web services usage through a log based architecture," in *Web Services, 2004. Proceedings. IEEE International Conference on*, 2004, pp. 61-69.

[17]   W. Wang and T. E. Daniels, "Building evidence graphs for network forensics analysis," in *Computer Security Applications Conference, 21st Annual*, 2005, p. 11 pp.

[18]   K. Shanmugasundaram, N. Memon, A. Savant, and H. Bronnimann, "ForNet: A Distributed Forensics Network," in *Proceedings of the Second International Workshop on Mathematical Methods, Models and Architectures for Computer Networks Security*, St. Petersburg, Russia, 2003.

ISBN 9781503217744